TRIASSIC NEW MEXICO

DAWN OF THE DINOSAURS

NEW MEXICO MUSEUM OF NATURAL HISTORY & SCIENCE

New Mexico Museum of Natural History and Science
A Division of the New Mexico Department of Cultural Affairs
2008

Text by Spencer G. Lucas and Selena Connealy
Artwork by Matt Celeskey and Mary Sundstrom
Photographs by David Baccadutre
Design and layout by Matt Celeskey

Front Cover:
NMMNH P-42200 *Coelophysis bauri* skull and cervical vertebrae
Photograph by David Baccadutre

Frontispiece:
Late Triassic forest with aetosaur and pterosaurs
Illustration by Mary Sundstrom

Published by New Mexico Museum of Natural History and Science
1801 Mountain Road N.W. Albuquerque New Mexico 87104 USA
http://www.nmnaturalhistory.org

Original printing: April, 2008
Printed in Canada

ISBN 13: 978-0-615-18330-5

Dedication

Mary Gavin had the vision and the energy to work towards creating a museum where the people of New Mexico could learn about their state's natural history and further their understanding of science. Because of Mary, and those other far-sighted people she worked with, the New Mexico Museum of Natural History and Science now provides educational experiences and promotes scientific inquiry through focused collections, research, public programs and exhibitions.

Mary Gavin's dedication helped to create the New Mexico Museum of Natural History and Science, and the Museum dedicates its Triassic exhibit hall to Mary and her family.

This book is dedicated to Mary.

The Certified New Mexico Fossil seal, used in this book and on labels throughout the Museum, indicates that you are looking at real fossils found in New Mexico.

The Exhibit

In May 2008, the New Mexico Museum of Natural History and Science opened its Triassic exhibition (*Dawn of the Dinosaurs: Triassic New Mexico*) in the 2,300 square foot Edward and Mary Gavin Family Dawn of the Dinosaurs Hall. The exhibition displays and interprets numerous Triassic fossils discovered in New Mexico that are part of the Museum's permanent collections. It teaches a basic concept of biodiversity—at any time in earth history, organisms belong to one of three groups: those that are first evolving, those that are becoming extinct, and those that are persisting. The Triassic provides an excellent example of this concept, with dinosaurs and mammals first evolving during the Triassic, many large amphibians and reptiles suffering extinction, and the persistence of organisms such as lungfish and conifers.

Triassic Time

The Triassic Period of earth history lasted about 51 million years, from 200 to 251 million years ago. It was the first period of the Mesozoic Era, the "Age of Dinosaurs," which encompasses the Triassic, Jurassic, and Cretaceous.

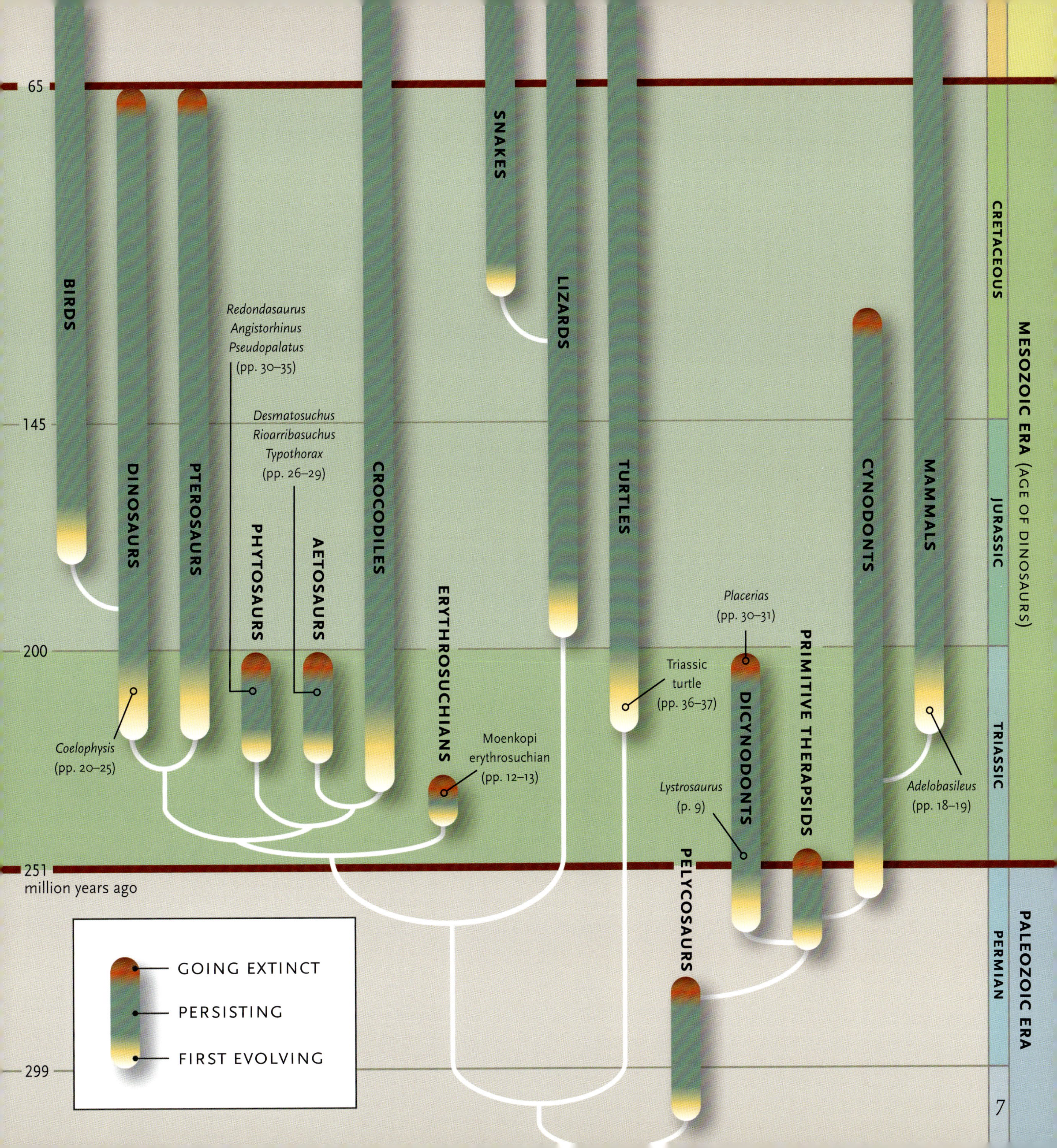
65
145
200
251
million years ago
299
BIRDS
DINOSAURS
PTEROSAURS
PHYTOSAURS
AETOSAURS
CROCODILES
ERYTHROSUCHIANS
SNAKES
LIZARDS
TURTLES
DICYNODONTS
PRIMITIVE THERAPSIDS
PELYCOSAURS
CYNODONTS
MAMMALS
Redondasaurus
Angistorhinus
Pseudopalatus
(pp. 30–35)
Desmatosuchus
Rioarribasuchus
Typothorax
(pp. 26–29)
Coelophysis
(pp. 20–25)
Moenkopi
erythrosuchian
(pp. 12–13)
Triassic
turtle
(pp. 36–37)
Placerias
(pp. 30–31)
Lystrosaurus
(p. 9)
Adelobasileus
(pp. 18–19)
CRETACEOUS
JURASSIC
TRIASSIC
PERMIAN
MESOZOIC ERA (AGE OF DINOSAURS)
PALEOZOIC ERA
GOING EXTINCT
PERSISTING
FIRST EVOLVING

The Triassic World

During the Triassic Period, all the continents were united into one supercontinent called Pangea (pan • JEE • uh). A single, vast ocean called Panthalassa (pan • thuh • LASS • uh) surrounded Pangea. What is now New Mexico was located in western Pangea just north of the Equator.

The Triassic world lacked ice caps, and warm and subtropical climates prevailed across Pangea. There were only two seasons, wet and dry.

The largest mass extinction in the history of life happened 251 million years ago, at the end of the Permian Period (beginning of the Triassic). Paleontologists estimate that 90% of marine species were eliminated. Seafloor communities that had been dominated by brachiopods, crinoids, rugose corals, and bryozoans were replaced by communities dominated by mollusks—including ammonites, bivalves, and gastropods. The end-Permian extinction coincided with the eruption of the Siberian traps, one of the greatest volcanic events in Earth history. The climatic effects of this volcanism may have caused the extinction.

The most compelling evidence for the supercontinent Pangea is the discovery of the fossils of a small reptile named *Lystrosaurus* (LIST • row • sore • us). Fossils of *Lystrosaurus* have been found on four continents, Africa, Antarctica, Asia, and Europe. How did an animal that could not swim or fly end up in these places now separated by vast distances and oceans?

A land-living plant eater, *Lystrosaurus* must have been able to walk between these now distant locations. The location of the fossils are evidence that the Early Triassic continents were united in a single landmass (Pangea).

New Mexico has an extensive Late Triassic rock and fossil record.

There are no Early Triassic rocks in New Mexico, nor are there rocks from most of the Middle Triassic. During that time span, Triassic New Mexico must have been an erosive landscape in which no rocks were formed.

Triassic rock layers in New Mexico record part of the Middle Triassic (Moenkopi Formation) and all of the Late Triassic (Chinle Group) history of the state. These rocks are red sandstones, conglomerates, and shales that were deposited by river channels, on floodplains, or in lakes under a hot, subtropical climate.

Triassic outcrops in New Mexico

The Moenkopi Formation

In northern New Mexico, the Moenkopi Formation is gray-red sandstone, conglomerate, and shale deposited by rivers that flowed to the north and northwest. A few fossil plants and reptiles, and many fossils of amphibians have been collected from New Mexico's Moenkopi rocks. They give us our first glimpse of the Triassic in New Mexico, a land of heavily vegetated river floodplains in which amphibian predators dominated the waterways. Large reptiles were the top predators on land prior to the appearance of dinosaurs.

Middle Triassic New Mexico

Eocyclotosaurus (ee • oh • sye • KLO • tow • sore • us) was an amphibian with a thick, flat and shovel-shaped skull found in the Moenkopi Formation in New Mexico. This fish eater is closely related to metoposaurs, a group of Late Triassic amphibians.

The only Middle Triassic fossil plants known from New Mexico are leaf impressions of the cycad *Zamites* (za • MIGHT • eez), which grew in thickets along the Moenkopi river banks.

Zamites powelli
Leaf impression, NMMNH P-33279
Moenkopi Formation
Guadalupe County, New Mexico

◀ An incomplete skeleton of an erythrosuchian (er • RITH • row • suke • ee • an), a large predatory reptile, is one of the rare reptile fossils known from the Moenkopi Formation in New Mexico. Erythrosuchians probably hunted the amphibians that are so abundant as fossils in the Moenkopi Formation.

Triassic Forests

In the Four Corners states of Arizona, Utah, Colorado, and New Mexico, rocks of Late Triassic age (200 to 225 million years old) contain many fossil logs. The most famous of these logs are found in the Petrified Forest National Park of northern Arizona.

Fossilized wood, sometimes called petrified wood, is formed when the original woody material is replaced by the mineral silica (silicon dioxide), making the fossils extremely hard. The bright colors come from impurities of iron, manganese, copper, and other elements.

Araucarioxylon arizonicum
Fossilized wood, NMMNH P-20382
Petrified Forest Formation, Chinle Group
Navajo County, Arizona

Living horsetails have jointed stems and whorls of thin leaves. Triassic horsetails, called *Neocalamites* (KNEE • oh • cal • uh • MIGHT • eez), were similar but generally larger. All horsetails are sphenophytes, a group of plants related to ferns.

The stems of the extinct horsetail formed branches that extended out and up from the central trunk. Long, thin oval-shaped leaves were attached to the stems, typically forming round whorls of many leaves.

Sanmiguelia lewisii
Leaf impression, NMMNH P-49570
Garita Creek Formation, Chinle Group
San Miguel County, New Mexico

Some paleontologists identify the Late Triassic plant *Sanmiguelia* (san • me • GUELL • ee •uh) as a palm, which, if correct, would make it the oldest flowering plant (angiosperm). Most paleontologists, however, regard *Sanmiguelia* as a bizarre plant whose evolutionary relationships to other plants are unclear.

Lungfish fossils date back to the Devonian Period and document about 400 million years of evolution.

Lungfish toothplates

Instead of individual teeth, lungfish have enamel-covered plates. These Triassic toothplates are very similar to the toothplates of a living lungfish.

(this page)
ARGANODUS (ar • guh • NO • dus)
Toothplates, NMMNH P-43979, P-43981
Petrified Forest Formation, Chinle Group
Navajo County, Arizona

(facing page)
Lungfish Burrow, NMMNH P-50409
Redonda Formation, Chinle Group
Quay County, New Mexico

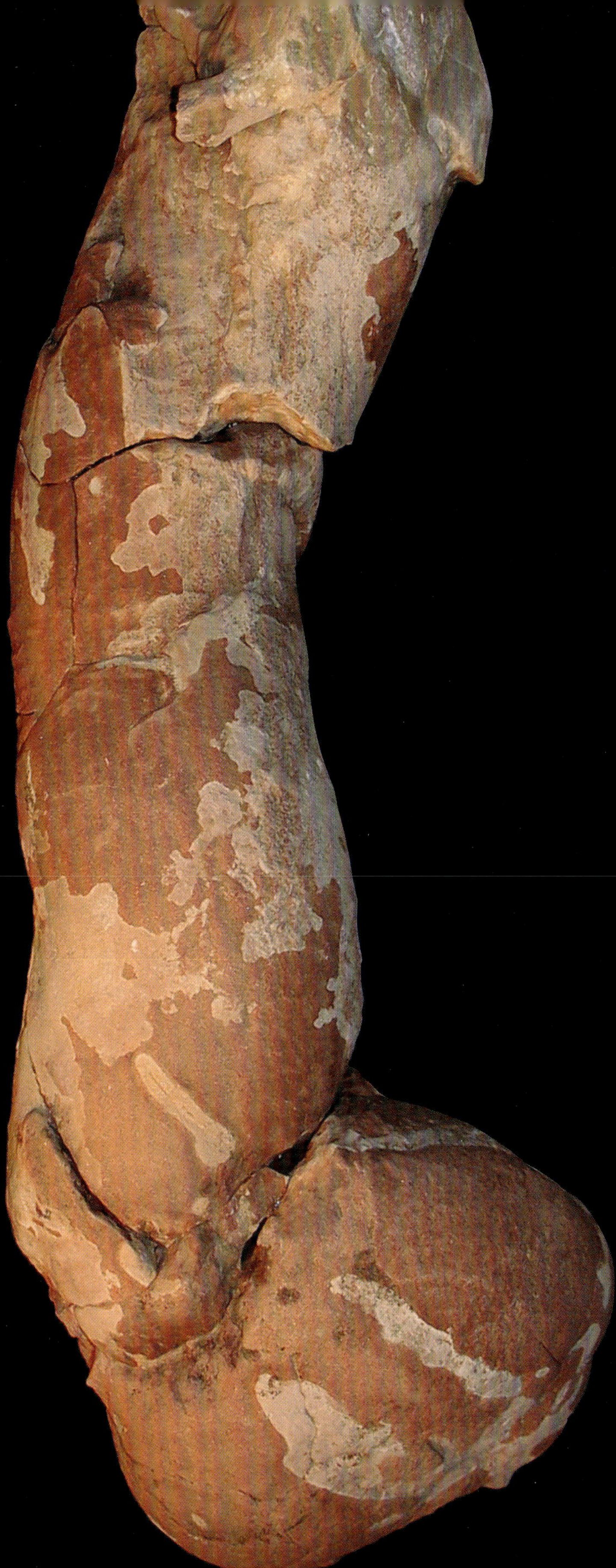

Today, some lungfish burrow into lake or river bottom mud to wait out dry spells (estivate). Some Triassic lungfish did the same, and casts of their burrows are found as fossils. Notice the slightly helical shape of this fossil burrow and the expanded, bulb-like ending. These are characteristic features of the estivation burrows of living and extinct lungfishes.

The oldest mammal fossils and the oldest dinosaur fossils are found in Late Triassic rocks; paleontologists consider both groups to have originated nearly simultaneously. Dinosaurs soon became the dominant land animals of the Jurassic and Cretaceous Periods. Mammals remained relatively small during the age of dinosaurs and didn't dominate until the Cenozoic Era, starting about 65 million years ago.

Actual size

Adelobasileus, the oldest mammal

We know mammals today by features that generally do not fossilize; they are hairy vertebrates that give live birth, and the females produce milk to feed their young. However, paleontologists identify mammals by their unique skeletal characteristics, including the presence of three bones (auditory ossicles) in the skull for hearing and the fact that their lower jaw is made up of a single bone.

This is the skull of the oldest known mammal, *Adelobasileus* (uh • dell • oh • ba • SILL • ee • us), a tiny mouse-sized inhabitant of the Late Triassic forests. Discovered by NMMNHS paleontologists in 1989, the skeletal structure of the ear region identifies the skull as mammalian.

Adeolbasileus cromptoni
Partial skull, NMMNH P-12971
Tecovas Formation, Chinle Group
Crosby County, Texas

Coelophysis

About 2 meters (6.5 feet) long, lightly built and bipedal (walking on hind legs), *Coelophysis* (see • low • FY • sis) was a quick and agile predator of the Late Triassic floodplain. The long curved neck, relatively large eyes, and grasping hands suggest that the dinosaur was a visual hunter that grabbed its prey (insects, small reptiles, and mammals) with its mouth or hands.

This block of rock is full of skeletons of *Coelophysis*, a Late Triassic dinosaur that is New Mexico's official state fossil. These skeletons are part of an extensive and world-famous bone bed of *Coelophysis* discovered during the 1940s at Ghost Ranch near Abiquiu in Rio Arriba County, New Mexico.

COELOPHYSIS BAURI
Multiple skeletons
Rock Point Formation, Chinle Group
Rio Arriba County, New Mexico

Coelophysis bauri
Multiple skeletons
Rock Point Formation, Chinle Group
Rio Arriba County, New Mexico

This page:
(Left) Articulated skeleton

(Top right) Forearm and hand, ribs, other elements

Facing page:
(Top left) Skull with hyoid bones

(Middle left) Tail vertebrae

(Bottom left) Furcula (wishbone)

(Bottom right) Multiple elements

(Top right) Large snout and jaws

P.46615

Desmatosuchus haplocerus
Armor Plate, NMMNH P-22412
Santa Rosa Formation, Chinle Group
Santa Fe County, New Mexico

Rioarribasuchus chamaensis
Armor Plate, NMMNH P-50141
Petrified Forest Formation, Chinle Group
Rio Arriba County, New Mexico

Aetosaurs (ae • EE • tow • sores) were heavily armored reptiles. One of the most successful plant-eating reptile groups of the Late Triassic, aetosaur fossils are known from North America, South America, Europe, India, North Africa, and Madagascar.

Paleontologists distinguish different kinds of aetosaurs by their different kinds of armor. Armor characteristics include spikes, shape, and texture. Each distinctive kind of armor is thought to belong to a different species of aetosaur.

Typothorax coccinarum
Armor Plate, NMMNH P-51304
Petrified Forest Formation, Chinle Group
Rio Arriba County, New Mexico

(following pages)
Typothorax coccinarum
Skeleton, NMMNH P-12964
Bull Canyon Formation, Chinle Group
Quay County, New Mexico

A nearly complete skeleton of the aetosaur *Typothorax* was excavated by NMMNHS paleontologists from Upper Triassic rocks near Tucumcari, New Mexico. The fossil preserves the animal's skull (pictured at left), hind limbs, and arrangment of the armor plates in great detail.

About 225 million years ago, a phytosaur chewed on the end of this femur (thigh bone) of a dicynodont. The phytosaur tooth marks are circular to elliptical holes in the fossil bone.

The synapsid reptile *Placerias* (plu • SAYR • ee • us) was one of the largest herbivores of the Late Triassic. Herds of *Placerias* browsed on the Triassic floodplains.

Dinosaurs were not the top predators of the Triassic landscape. Instead, the crocodile-like phytosaurs (FY • toe • saurs) were the largest (up to 10 meters/32.8 feet long) and most ferocious meat-eaters on the floodplain. Phytosaurs probably attacked their prey by ambushing them with a sudden lunge from the water.

(inset)
Heterodontichnites hunti
Tooth marks on dicynodont femur, NMMNH P-13001
Santa Rosa Formation, Chinle Group
Santa Fe County, New Mexico

(top)
Placerias hesternus
Cast of skeleton, NMMNH C-1858
Bluewater Creek Formation, Chinle Group
Apache County, Arizona

EXIT

ANGISTORHINUS ALTICEPHALUS
Skull, NMMNH P-33527
Colorado City Formation, Chinle Group
Howard County, Texas

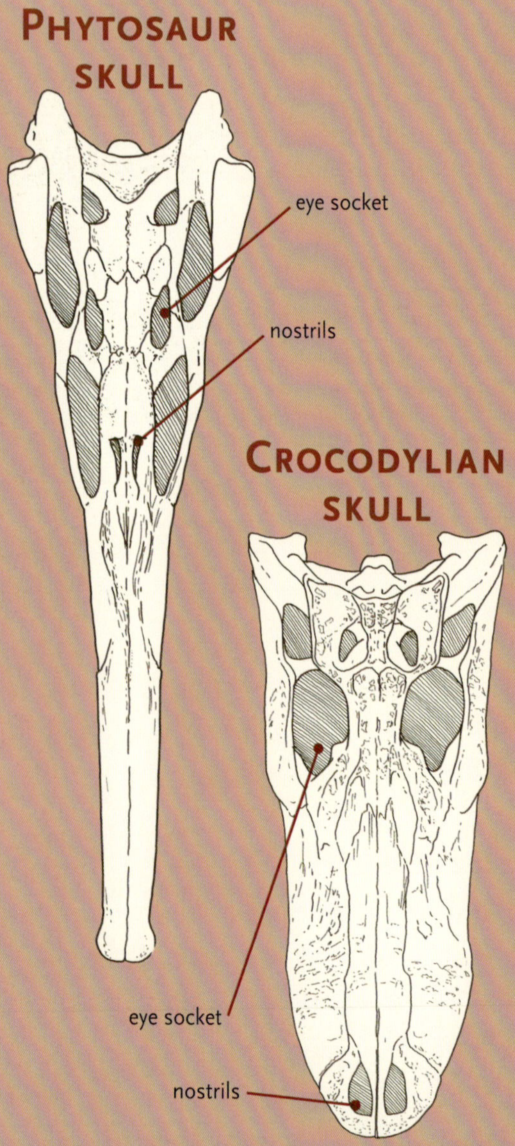

Phytosaurs thrived in the Late Triassic. Their fossils are found in North America, South America, Europe, North Africa, Madagascar, and India. Phytosaurs rapidly evolved into many species during the 25 million years of the Late Triassic.

Phytosaurs look remarkably like crocodiles, but the groups of reptiles are only distantly related to each other. The easiest way to distinguish a phytosaur from a crocodile is the location of the nostrils. On the crocodile, they are at the tip of the snout, but on a phytosaur they are well back on the snout, near the eyes.

The similarities between crocodiles and phytosaurs are an example of convergent evolution. Although both phytosaurs and crocodiles are reptiles, they are only very distantly related. Their shared characteristics derive from both organisms being adapted to similar environments, not from shared ancestry.

Boys and Girls

Sexual dimorphism—anatomical differences between males and females of a single species—is common among living reptiles. These phytosaur skulls from a bone bed in northern New Mexico appear to be from a single species. However, they show two different kinds of nostril anatomy—a very large-crested nostril and much more slender-crested nostril. These may be sexual dimorphs, but paleontologists are not certain which nostril anatomy is male and which is female.

Pseudopalatus buceros
Skulls, NMMNH P-31292, P-50040
Petrified Forest Formation, Chinle Group
Rio Arriba County, New Mexico

Turtles first evolved during the Late Triassic as reptiles with an armor shell, but unlike modern turtles, they could not pull their heads back into their shells. Triassic turtle shells are very similar to the shells of modern turtles.

North America's Oldest Fossil Turtle

Unearthed by NMMNHS field crews from the Upper Triassic badlands near Tucumcari, New Mexico, this newly discovered fossil is North America's oldest known turtle. The fossil consists of pieces of the upper shell (carapace) and the lower shell (plastron), as well as many pieces of spike-like armor that covered the turtle's neck and tail. The vertebrae are fused to the carapace, a characteristic of all turtles.

This Triassic turtle has a very thin shell which may indicate that it was amphibious. Other Triassic turtles, which have thick and heavily-armored shells, were land-living animals.

TRIASSIC TURTLE
Incomplete shell, armor, and limb elements,
NMMNH P-16697, P-45555
Bull Canyon Formation, Chinle Group
Quay County, New Mexico

TRILOPHOSAURUS JACOBSI
Juvenile skull and jaw, NMMNH P-41400
Trujillo Formation, Chinle Group
Borden County, Texas

The Triassic reptile *Trilophosaurus* (try • LOW • fo • sore • us) looks very much like a modern iguana, but has a completely different ancestry. The similarities between *Trilophosaurus* and iguanas are an example of convergent evolution. Their shared characteristics derive from both organisms evolving to survive in similar enviroments, not from shared ancestry.

*A*mphibians nearly went extinct at the end of the Triassic. The shovel-headed amphibians that had been among the most common land animals since the Devonian almost disappeared, with only a few species surviving into the Jurassic.

Buettneria perfecta
Skull, NMMNH P-36228
Juvenile skull, NMMNH P-36229
Colorado City Formation, Chinle Group
Howard County, Texas
Fossils collected and donated by Robert Kahle

Triassic to Jurassic

About 200 million years ago, major extinctions marked the end of the Triassic Period. On land, many of the Triassic amphibians and reptiles disappeared, including the metoposaurs, phytosaurs, and aetosaurs. In the sea, ammonites almost became extinct, and extinctions produced dramatic changes in some of the ocean plankton.

One likely cause of these extinctions was the massive volcanic eruptions that took place as the supercontinent Pangea began to break apart at the end of the Triassic. The volcanoes injected huge amounts of carbon dioxide, sulfur dioxide, and other gases into the atmosphere. World climate changed. Ocean chemistry was altered. Most groups of plants and animals were not severely affected by these changes and persisted to populate the Jurassic world. Other groups could not adapt and became extinct.

Ammonites nearly went extinct at the end of the Triassic. Only one genus survived into the Jurassic, giving rise to the diversity of ammonites that swam the Jurassic and Cretaceous seas. Triassic ammonites were diverse and most had complex shell shapes and ornamentation, unlike the smooth-shelled ammonite (*Psiloceras*) that survived the extinction.

Ammonite Shells

Middle Triassic through Early Jurassic

Churchill, Pershing, Mineral, and Nye Counties, Nevada

Fossils collected and donated by Jim Jenks

Trachyceras desatoyense NMMNH P-51290

Proarcestes gabbi NMMNH P-51291

Tropites crassicostatus NMMNH P-51287

Psiloceras pacificum NMMNH P-51297

Silberlingites cricki NMMNH P-51300

Acrochordiceras carolinae NMMNH P-51303

Favretoceras wallacei NMMNH P-51294

Tropigastrites lahontanus NMMNH P-51299

Jurassic
Triassic

Grallator
Footprint, NMMNH P-44187
Redonda Formation, Chinle Group
Quay County, New Mexico

The oldest definite dinosaur tracks are called *Grallator* (GRALL • ay • tor). They are the tracks of relatively small, meat-eating dinosaurs like *Coelophysis*, that lived during the Late Triassic.

Trace fossils like tracks and tooth impressions have different names than the animals that made them. Unless an animal literally "died in its tracks," paleontologists can't be certain whose foot made the impression. Because of this, the trace fossils receive their own names so that they can be discussed by scientists.

Further Reading

Fraser, N. 2006. *Dawn of the dinosaurs: Life in the Triassic.* Indianapolis, Indiana University Press, 307 pp.

Harris, J. D., Lucas, S. G., Spielmann, J. A., Lockley, M. G., Milner, A. R. C. and Kirkland, J. I. (editors). 2006. *The Triassic-Jurassic terrestrial transition.* New Mexico Museum of Natural History and Science, Bulletin 37, 607 pp.

Heckert, A. B. and Lucas, S. G. (editors). 2002. *Upper Triassic stratigraphy and paleontology.* New Mexico Museum of Natural History and Science, Bulletin 21, 301 pp.

Long, R. A. and Houck, R. 1988. *Dawn of the dinosaurs: The Triassic in Petrified Forest.* Petrified Forest Museum Association, Petrified Forest National Park, 96 pp.

Lucas, S. G. and Hunt, A. P. (editors). 1989. *Dawn of the age of dinosaurs in the American Southwest.* Albuquerque, New Mexico Museum of Natural History, 414 pp.
Available online at http://www.nmnaturalhistory.org/sci_dinodawn.html

The New Mexico Museum of Natural History and Science has assembled a comprehensive look at New Mexico's State Fossil, the Triassic dinosaur *Coelophysis*, online at http://www.nmstatefossil.org

More information on the Triassic in New Mexico, and the *Dawn of the Dinosaurs* exhibit, can be found at the New Mexico Museum of Natural History and Science website: http://www.nmnaturalhistory.org

Index

Acrochordiceras 40
Adelobasileus 7, 18–19
aetosaurs 7, 26–29, 40
ammonites 40–41
amphibians 13, 39
Angistorhinus 34
Araucarioxylon 14

Buettneria 39

Chinle Group 10–11
conifers 14
Coelophysis 7, 20–25, 42
convergence 34, 38

Desmatosuchus 26
dicynodonts 7, 9, 30–31
dinosaurs 7, 20–25, 42

Eocyclotosaurus 13
erythrosuchian 7, 12–13

Favreticeras 41
flowering plants 15

horsetails 15

Grallator 42

Heterodontichnites 30

Jurassic Period 7, 40–41

Late Triassic 11, 14–42
lungfish 16–17
Lystrosaurus 7, 9

mammals 7, 18–19
metoposaurs 13, 39, 40
Middle Triassic 11, 12–13
Moenkopi Formation 11, 12–13

Neocalamites 15

Pangea 8–9, 40
Permian Period 7, 9
petrified wood 14
phytosaurs 7, 30–35, 40
Placerias 7, 30–31
plants 13, 14–15
Proarcestes 40
Pseudopalatus 35
Psiloceras 40–41

Redondasaurus 7, 31–33
Rioarribasuchus 26

Sanmiguelia 15
Silberlingites 41

trace fossils 17, 30, 42–43
Trachyceras 40
Trilophosaurus 38
Tropigastrites 41
Tropites 41
turtles 7, 36–37
Typothorax 27–29

vulcanism 9, 40

Zamites 13

Exhibition Credits

Funding for the development, construction, and promotion of *Dawn of the Dinosaurs: Triassic New Mexico* was provided by Governor Bill Richardson, the New Mexico State Legislature, and the following:

The Jane Sandoval Endowment
Upjohn Stearns Trust Fund
Gae E. & Lars Eisenhardt

Thomas K. Reed
Jon D'Andrea

Ann Gavin Cruikshanks
August Wood Cruikshanks
Florence H. Dapples
The Ettinger Foundation
William M. B. Gavin
Herman Mauney
H. W. & Maryann Nordyke
Jacquenette C. Ostheimer
John L. & Donna Rust
Robert Stamm
W. John & Dorothy J. Swartz

Richard & Margaret Bice
Erick F. Carlgren, DDS
William & Jenny Cowles
James H. Duncan
Richard & Claire M. Kropschot
Betty McElvain
Tim H. McElvain
John J. & Janie A. Moseley

James Alley
William W. & Sandra K. Crawford
Suzanne Epps
Lloyd & Beverly Fadrigue
Joseph A. & Dora Fidel
James E. Loucks
Carolyn R. Mercer
Abe Silver

Promotional support provided by Sandia National Laboratories

Thanks also to the collectors who donated some of the fossils on display.

The exhibit would not have been possible without the dedicated efforts of the staff and volunteers at the New Mexico Museum of Natural History and Science and the New Mexico Museum of Natural History Foundation.

Picture Credits

Cover
Photograph by David Baccadutre

Pages 1–2
Illustration by Mary Sundstrom

Page 5
Photograph by Jim Thompson

Page 7
Diagram by Matt Celeskey

Page 8
Illustration by Mary Sundstrom

Page 9
Illustration by Matt Celeskey

Page 10
Photograph by Spencer G. Lucas

Page 11
Background illustration by Mary Sundstrom
Map after Heckert, A. B., Lucas, S. G., and Sullivan, R. M., 2000, Triassic Dinosaurs in New Mexico: New Mexico Museum of Natural History and Science, Bulletin 17. pp. 17–26.

Page 12
Photograph by Spencer G. Lucas

Page 13
Illustrations by Matt Celeskey
Photograph by David Baccadutre

Page 14
Araucarioxylon photograph by David Baccadutre
Petrified Forest National Park
Photograph by Spencer G. Lucas

Page 15
Illustrations by Mary Sundstrom
Photograph by David Baccadutre

Page 16
Illustration by Mary Sundstrom
Photograph by David Baccadutre

Page 17
Illustration by Mary Sundstrom
Photograph by Christie Gross

Page 18
Illustration by Mary Sundstrom

Page 19
Photograph by David Baccadutre

Page 20
Illustration by Matt Celeskey

Page 21
Photograph by Lorrie Latham

Pages 22–23
Photographs by David Baccadutre

Pages 24–25
Illustration by Mary Sundstrom and Matt Celeskey

Page 26–27
Illustrations by Matt Celeskey
Photographs by David Baccadutre

Pages 28–29
Photographs by Spencer G. Lucas and Mike Pierce

Page 30
Placerias photograph by Matt Celeskey
Heterodontichnites photograph by David Baccadutre

Page 31
Illustration by Matt Celeskey and Mary Sundstrom

Pages 32–33
Mike Pierce working on the skeleton of *Redondasaurus*, photograph by David Baccadutre

Page 34
Photograph by David Baccadutre
Diagram by Mary Sundstrom

Page 35
Photographs by David Baccadutre
Illustrations by Matt Celeskey

Page 36
Illustration by Mary Sundstrom

Page 37
Photographs by David Baccadutre

Page 38
Illustration by Mary Sundstrom
Photograph by David Baccadutre

Page 39
Photograph by David Baccadutre
Illustration by Matt Celeskey

Pages 40–41
Photographs by David Baccadutre

Page 42
Photograph by David Baccadutre

Page 43
Illustration by Mary Sundstrom and Matt Celeskey

Page 48
Illustration by Mary Sundstrom

We would like to acknowledge the efforts of Christie Gross, who digitized and prepared many of these images for publication.

Adelobasileus cromptoni, the first mammal
Actual size